Roads and Tunnels

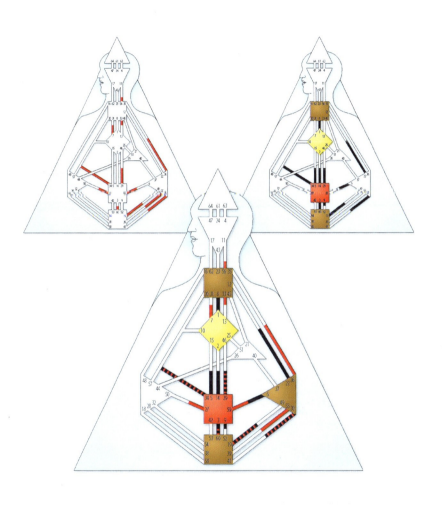

RA URU HU

Roads and Tunnels

The real secret of Human Design is the simple. It begins with the simple and it ends with the simple. The unique element of Human Design is its binary calculation. It is the very basis upon which we look at or create a Human Design rave chart. It's based on two points in time. Each of these points in time act as imprinting agents, and each of those imprints carry their own value, carry their own energy, carry their own purpose. And inherently have no relationship to each other, other than being juxtaposed within the rave chart, and within the rave chart come to grips with the presence of the other.

The calculation begins with the time and place of birth. It is this unique moment of time and place of birth that allows us to be able to find our secondary calculation. The secondary calculation is 88 degrees of the Sun before one's birth. That is, 88 degrees of the movement of the Sun. It is approximately 88 or 89 days before one is born.

> The real secret of Human Design is the simple.

The calculation at the time of birth is made when the body completely leaves the womb. It's not about cutting the umbilical cord; it is not about the first breath. It is about the moment of being in an entirely separate space. And it is this moment that we calculate for the position of the planets and the position of the nodes. It's this moment that we calculate for the personality.

Remember, in the context of Human Design, in the cosmology of Human Design, we're endowed with crystals of consciousness. We have a personality crystal that sits in the crown head chakra, the head center. And this personality crystal is an aspect of the original yang. It manifests who we think we are. And it's one of the keys in Design to recognize that the personality lives in an illusion of what it thinks it is. The illusion is not so much what it is in contact with and concludes that is itself, the illusion is that's all there is to the self, and everything else is confusing.

The personality is the passenger in the back seat of a car; it's not the vehicle. The vehicle is the by-product of the design crystal. That is, the design crystal manifests, or operates, our bio-genetic vehicle. And it is the design crystal in which the information of the design crystal is the calculation that is pre-natal, the calculation that is made 88 degrees of the movement of the Sun before birth.

It is this design information that is really the great revelation of Human Design. It allows us to come into contact with a hidden aspect of our nature. It allows us to see, in fact, who we truly are. And what it allows us to see is what operates within us at a purely mechanical level that does not have any contact with the personality. In other words, does not become part of our self-reflected awareness. In fact, it can be very, very confusing. Who we think we are, can be desperately frustrated with what the body demands it must be.

These two calculations are the basis of the binary in Human Design. They operate differently. And not only do they operate differently in that the personality is something that we all can have access to, and the design is something that truly is hidden from us. But their impact on our nature, once we integrate them into the body graph, is entirely different. It is very different to have a personality that speaks, and for a being to be consciously aware of that, than to be a being whose unconscious speaks, and not consciously understanding why or how all of that is taking place.

The same functions within the same channel brought by different aspects in the calculation result in a totally different phenomenon in terms of experiencing what it is to be that life impacted in that way. And so much of the deep, deep fear that we have as human beings is because we do not understand this hidden aspect of our nature, and why it operates the way it does, and why so often it's in conflict with who we actually think we are.

The personality is what we most identify with. It is all we're really connected to. Let's look at our example. Our example to begin with is a generator. We have a defined sacral center. And the moment that we have a defined sacral center in the chart we know that somebody is a generator, we know that the key in their life is their strategy of responding and non-initiating. This is a pure generator. The sacral center is connected down to the root center, and we can see that that is a personality connection. In other words, this is somebody who has identified with their power. This is somebody that when they think of themselves will recognize that there is all this energy within them.

> The personality is what we most identify with.

By the way, this is a recipe for deep frustration and depression. This design is a split definition. There are two areas of definition, and they're not connected to each other. The throat is connected by definition to the G center. Again, this is a personality connection. So here is somebody who's strongly identified with their voice. Yet, their voice speaks of the need, the deep need within their identity to express themselves and to make a contribution through that expression.

So often the verbal gunslinger, somebody whose throat is defined but not connected to a motor, ends up making commitments, ends up stepping into things, ends up offering themselves to things, ends up saying they will do things that ultimately they cannot do because they do not have a consistent access to their energy, to the sacral and its power. And the sacral center when it's not operating correctly only leads to frustration: the frustration of not being able to use one's energy in a healthy way, the frustration that comes with feeling all of that within one, and being forced into activities that are simply not satisfying.

Here is somebody that identifies with their need to make a contribution, identifies with the reality of the power that lies within them, and then is at a complete loss to figure out why it is they cannot do what they say they would like to do, or what they think they would like to do, that they cannot make the contribution that they want to want to make when they want to make it. It's not that it never happens.

Anyone who has a split definition is going to spend a great deal of their life looking for the right connections; looking for the right people to make the bridges that allow them to feel whole in the expression of their potential.

But our focus here is on the identification, the roads, in which this being is going to see itself, the limitation of the personality only to be identified not simply with a half of us, because it's not. It may be the duality of the data base, but it is an aspect of a whole that's greater than what it is. The easiest way to see that is to remember what this complete design looks like, and to recognize what's missing.

The moment that you re-integrate the personality back into the totality, back into the whole that's greater than the sum of its parts, you come to a defined emotional system. The solar plexus center, and a very, very powerful solar plexus center with all of these aspects pointing out of the solar plexus center, within the context of the whole design is defined.

This is somebody who is an emotional generator. This is somebody who has to live by the authority of the emotional system that states that there is no truth in the now, that one must wait. That it's not simply a question of being a generator, ready to respond, but a generator that's ready to respond in time. And only in time, and only over time, in order to find clarity.

Yet, we're dealing with a personality who we think we are. And that personality is not going to be identified with being emotional. It's not there within the construct of the personality. It's not there for the personality to identify with. And it means that they do not identify with waiting and they do not identify with the emotional wave. They barely know that it's there. And only have to come to grips with the possibility that they may, in fact, be emotional, when they're confronted with it over and over again in their life process.

Yet, it's not a simple thing for them to be able to accept it. And it's not stubbornness. And it's not simply a matter of being provocative. The fact of the matter is that because the personality does not connect to it, the personality does not identify with it, doesn't recognize it.

Again, this is the magic in Human Design. You see, we all recognize these unconscious aspects in ourselves. We recognize them as traits that we come to grips with over time. We don't know why they're there; we don't know why they work the way they do. But they keep on seeming to be there as part of our nature, and people keep on reminding us, so it must be there.

So when we're dealing with the personality and we finally have an opportunity to allow it to see how the other side actually operates, there's a wonderful chance to let go. And to let go of the burden of the fear that comes with not

> There is a major difference between those gates that are active and functioning, and those gates that are dormant.

5

being able to accept your whole self.

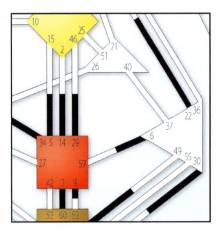

In looking at the identification of the personality, the first thing to recognize is that there is a major difference between those gates that are active and functioning, and those gates that are dormant. In looking at the emotional system and the three gates that are coming out of the emotional system that are part of the personality calculation, the 6th gate of Conflict, the 22nd gate of Grace, and the 30th gate, the Clinging Fire, those three gates are not regularly, actively accessible to the personality.

> The focus of the personality is on those gates that are active and those gates that are part of the definition.

Now, obviously within the context of the design, that center is always functioning. But it's functioning through an unconscious bridge. There are times when they have access to that, and times when they do not, but it's not consistent. What is consistent for the personality, or better expressed, where the focus of the personality is, is on those gates that are active, those gates that are part of the definition, obviously. But also those gates, like you see in the sacral center in which you have six gates through the sacral root system that are activated as part of that definition. All of that is part of the identification of the personality.

Think about the sacral center in this being. The sacral center is about our availability in this life. In other words, it's about what energy is there and how that energy is available, under what circumstances, under what conditions. It's highlighted by the presence of the 29th gate; the 29th gate is the gate of making commitments. It's a gate of saying "yes." This person identifies with making commitments.

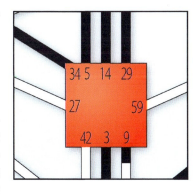

Remember, it's one of the constants of the generator's life that they're conditioned to be manifestors; they live as failed manifestors with its attendant frustrations, but keep on trying to succeed as manifestors regardless. In other words, they keep on initiating; they keep on trying to do things, rather than waiting, rather than responding, rather than not initiating.

Here is a generator with the 29th gate, and there is built into it this capacity to be ready to make commitments and it identifies with that. It identifies with the 14th gate, Possession in Great Measure, the need to find the right direction in order to be able to find wealth, in order to be able to find possessions, in order to be able to bring those things into the life that enrich the life.

Here's somebody who identifies with ritual and patterns through the 5th gate. They identify with the way in which they do things, and the time system that they're involved in. They identify with power. They identify with the empowerment that's possible through the 34th gate. And they identify with looking at the details, the 9th gate.

Here is somebody whose personality is deeply connected to the individual process. Both definitions are individual and individual definition brings the chemistry of melancholy. This chemistry of melancholy is very important to grasp, it is by any other name: the muse. In other words, this chemistry that comes as a sense of sadness, an aroma if you will, of sadness, is in fact a muse that is calling one. And it's essential for beings that have individual definition, strong individuality within the context of

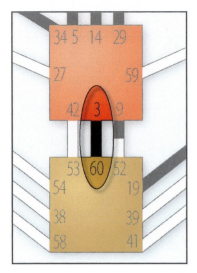

their design, to have some kind of creative outlet in order to move this melancholic energy.

In the 3/60, which is the format energy of the individual process, depression is possible. In other words, the moment that real mutation cannot take place, that real transformation cannot take place, this opens up this being to all kinds of difficulties and depression in their life process.

Remember, the sacral center when it is responding, isn't simply a dumb instrument. What it's doing is that it's acting as a medium for everything that is connected to it. When this being is asked something, or when this being responds to something, they're responding with that sacral and all of its connections, and most of all, the inherent quality of that definition.

> In the channel 3/60 (the format energy of the individual process) depression is possible.

What that definition is saying is that you say to this being "would you like to," and they go "ahunh," and what that "ahunh" is saying is that this is something that can be mutative. But the decision is being made within the context of the sacral and its many connections. This can be mutative. This is something where I can work on the details, this is something that could enrich me, this is something that fits into my pattern, and this is something I can commit to. In other words, all of the aspects that are there are all coming together in a synthesis to make that response.

It's deeply intelligent, if you will. It's simply not mental. It's genetically intelligent. And it is responding through that whole totality to anything that's there. Here is somebody that identifies with that. But of course, their difficulty lies in being the split definition. That the very contribution they identify with in the throat is not necessarily something they can do rooted in their sacral. And more than that, as we know, there is this hidden element; there is this emotional system that clouds the capacity of the sacral to see clearly.

> One of the deepest themes of Design knowledge is to offer you a mechanical way in which you can make reliable decisions.

Here is somebody that from the personality level doesn't really trust that sacral response. It's often incorrect; that is, incorrect in the moment. This is an emotional being, after all, operating in the wave despite what this aspect of their design looks like. So here is somebody that's deeply identified and can become very melancholic, can become very depressive if they don't find something through which they can be mutative, that they can make a commitment to, that they can fulfill. And that they're not going to be able to do that unless they're able to make a decision that they can trust.

One of the deepest themes of design knowledge is to offer you mechanically a way in which you can make reliable decisions. Because reliable decisions save the life, eliminate the burden, get rid of the blame and the shame and the guilt and all of that stuff. And it gets rid of the mind game that goes along with the decision making process that's always caught in the 'this and that.'

Reliable decisions are key. This being cannot make a reliable decision unless they're waiting out their emotional wave. And here is somebody

> The profile is an instrument of the Incarnation Cross.

whose personality refuses to identify with it because it's simply not there.

Profile is one of the essential views of Human Design. And in looking at profile, the personality profile represented by the lines of the personality Sun/personality Earth, we're dealing with a second line theme. This second line theme is a theme of being called in life. It's a theme of being left alone; "leave me alone and don't bother me". It's a theme of the hermit. This is the design of somebody who is a natural; this is a design of somebody that has real skills. But they're always subject to the projections of others, and they would rather be left alone.

The projection that immediately comes to somebody like this is that they can make a contribution and that they have the energy to do so. And yet, they themselves can find it exceedingly difficult and frustrating to hear that and to have that placed before them, only to be in a situation where it is so difficult for them to be able to do so unless they're functioning as themselves.

The profile is an instrument of our incarnation cross, our purpose in this life, the costume that we wear in this life in order to fulfill what our mechanics have offered us. Here with a second line personality that real purpose can only be fulfilled through waiting for the call. And not only in waiting for the call, but being reticent to accept any call in the moment. Ultimately, unless this being comes to a point in which they come to grips with the nature of their design and allow themselves the privilege of letting life come to them, they will live on the planet of suffering and they will have such a difficult time.

But you can see something. The real difficulty in dealing with somebody like this is that they do not identify with the most important aspect of their design in terms of being healthy and efficient. They simply don't. And it's not like you can bring them into therapy and counsel them and go through all of this process because they do not identify with it. They don't.

You see, philosophies are great, they're interesting, and they're things to chew on. They make for good table conversation. But they do not free the spirit. It's not like you can say to this being that they must become identified with their emotional nature.

It's not true. It's not about them getting into contact with something they cannot get into contact with. It's about bringing them to the mechanical and logical recognition of how they are structured, for them to see very clearly in the mechanism how they operate. And then to give them the only thing you can give them, their strategy to experiment with.

> What changes our life is when we follow the mechanics of our nature, and then there is no need for any effort.

You see, it's about the simple. What we can learn from our design is that we can learn to accept all of what we are. And yet, it is not conscious acceptance that will change our life. What changes our life is when we follow the mechanics of our nature, and then there is no need for any effort. There's no need to try to redeem; there's no need to try to heal. One can simply be what is correct.

It is this correctness that it is all about. Here is somebody that can only be correct when they're following the strategy of what it is to be an emotional generator—to play hard to get. One of the most important things that this view of roads that are operating in this being brings you to, is that you can bring this person to recognizing their limitation. And in recognizing their limitation, being open to the experiment, that really will bring them success.

They need to be called. And they need to identify with that call, and they can. It is the 14.2, Possession in Great Measure, and the 8.2, the Earth, the expression in the throat.

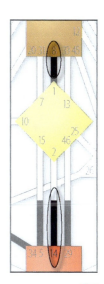

They can identify with being called. The moment that they're asked to speak, they already know how much more powerful it is for them than the moment that they try to speak. They've already experienced it in their life because they do have conscious identification with it; they've already experienced in their life, the benefit of being called. But they don't have the fearlessness to wait in between for the next call, that under this false pressure that they must, they plunge in and again go through the deep frustration and forget that being called is the way for them.

> when you're looking at somebody's design, the first thing to look at is the personality.

In returning to the rave as a whole, and in continuing our look at the profile, if you look on the design side, the Sun is in the 29.4, and the Earth is in the 30, fourth line. This cross is a right angle Cross of Contagion. The 30th gate, fourth line, is the line of Burnout. This is somebody that, without real guidance early in their life, is simply going to explode at the emotional level and can really simply burnout, be exhausted and fatigued, which is a fourth line theme, and the theme of the unconscious.

By being able to understand that when we're looking at somebody's design, that the first thing that we need to do is look at the personality. We need to look at the personality in the sense of looking to see what somebody is identified with, who they think of as themselves.

And in the moment that we see who they think of as themselves and its complexity, we can look and see what's missing. That is, the rest of their design that they do not identify with, and the problems that they can bring into their life process. And that in bringing somebody to their design, bringing them to the recognition that all that they see there in the black, that's just who they think they are. But it's not just what they are. There is so much more that has to be dealt with. And that it's in coming to grips with the other side that we come to grips with a deep,

deep, deep need within us to find peace—the mechanism of the unconscious.

Think how difficult it is for this being to have a sense of fulfilled purpose. The unconscious, that the unconscious is a fourth line theme.

Fourth Line Themes

Behavioral Identity	Opportunist
Projected Attitude	Abdicator
Limited Perspective	Fatigue
Aspired to Role	Aloneness
Bonding Strategy	Confident or Not
Security Strategy	Benefactor / Dependant
Emotional Resonance	Kindness / Meanness
Awareness Resonance	Corruption or Not

The fourth line theme is a theme of brotherhood and sisterhood. It's a theme of friendship; it's a theme of being communal and being open. It's about the opportunist; it's about all the opportunities that one can get in this life by having the right beings around them.

And yet in order to have the right beings around you, there has to be a part of you that's open to that in the first place. Here's somebody who is a 2/4, they're here to be called. And they naturally pull that call out by bringing people to them, socially, through their unconscious and its power. Yet, they do not identify with that.

And if you ask this second line personality if they are social and open, they will say no, that they would rather be on their own, that they would rather be left alone. Yet, it is one of the deep confusions in their life. Why is it that there are these people around me, what is it that they want from me, what is it? And they do not truly know.

And that the more you confront them with it and the more you say, but it is so, the more that they will deny that and simply see it as some kind of false projection. And it's only when they begin to understand that at the unconscious level they have a mechanism that draws people to them, that they can begin to see that they're designed to receive the call, and they're designed to be the opportunist waiting for the right call to come. And that the challenge of finding that right call is not about doing, and it's not about intellectualizing, it's about following the simple mechanic of one's nature.

It's about honoring, in this case, the emotional motor, honoring the fact that there is no truth in the now. That being an emotional generator is about playing hard to get in this life, that no one has easy access to this person. That no one can make them make a commitment easily, no one. They're here to be picky and fussy; they're here to wait for exactly the right opportunity and then to be fulfilled in that process.

You can only bring somebody with this design to that place if they're following their strategy, and you can see why. For most human beings it's very difficult to identify with all of those aspects of themselves that may be key in integral parts of their nature that they don't have conscious relationship to. That no matter what you tell them, and no matter how much you motivate them, doesn't mean it's going to work. What works is following your type and following your strategy.

The Design

It's a natural tendency for anyone that begins to study Human Design that whenever they're looking at a body graph what they're looking for automatically is definition. And yet at the same time in looking for that definition and seeing that the centers are colored in, they often forget to see the way in which they really are connected to each other.

One tends to assume certain things about a design based on the fact that there are a number of centers that are fixed. However, the very subtlety of the way in which we operate, that in looking at this design, how clear it is that the relationship between the sacral center and the root system and their connection to each other that's so powerful as a personality theme, that the moment that you try to get across to that emotional system you've got an unconscious gate. And that bridging loses its capacity to be something in which the personality can be identified with, despite the fact that there are personality activations that are there in the emotional system, and despite the fact that it's definition that is operating.

The unconscious—more than any other aspect of Human Design, it is this design calculation, this design information, that is so significant because it opens up an aspect to ourselves that has always really been hidden. The movement that took place at the end of the 19th Century and the beginning of the 20th Century through the work of Freud and Jung was this intuitive recognition of a much deeper aspect in the human experience that we don't really have clear access to, and the techniques developed to try to get access to this inner aspect, the names that were applied to it, the ideas and visions. Yet, in the simple mechanics in Human Design is the ability to

> The design information, that is so significant because it opens up an aspect to ourselves that has always really been hidden.

15

be able to see clearly how the so-called unconscious works.

I prefer tunnel, as in roads and tunnels, because they provide us with a metaphor that's really clear. The road is open, the road can be seen, we can experience it, we can experience what's going on on that road. We can identify with it. All of that becomes a part of our living experience.

> In looking at somebody's design we're entering into a very special place.

Yet, the tunnel is so different. We don't know what's in the tunnel. We don't know what direction anything is moving in, if there's anything in the tunnel at all. And we do not have any control over the timing of what's released through that tunnel, none, none whatsoever. It all becomes just a surprise. And more than a surprise, it becomes a way in which we're conditioned mentally to be insecure about ourselves in the world.

You see, out of these unconscious gates the forces that move us, the impact on various ways in which we handle life, they're always there, and yet we don't identify with it. So there is a discomfort within us. Why am I saying that, why am I doing this? I don't identify with that in my personality.

In the case of this being, why am I exploding? Why am I suddenly in an emotional crisis? I, not identified with being in an emotional crisis, am not identified with that at all. In other words, when we come to the design calculation we come to the most important aspect in terms of discovering who we really are.

Through our personalities we have a sense of identification. But here with the design, we truly need to have the mechanics of how that operates revealed so that we can be whole in that. It's not as if we don't have any access to design information. We do, but we do through hindsight, we do over time. This is our bio-genetic inheritance. It represents the traits that we've inherited genetically. And those genetic traits show themselves over and over. It is a genetic trait in this being to have a very powerful emotional system, and to respond emotionally in life. And whether they're identified with that or not, that is what's pouring out of the tunnel. And the tunnel is opening up to all of that emotional release.

This is the material that we're carrying. So over time in our lives, even though we don't identify directly with the way in which we're operating, we've learned to see that that is us. This being will come to a point where it says, yes, I don't know why, but I seem to have emotional outbursts, or emotional crises, or I seem to get into situations where I'm burnt out emotionally. In other words, over time we learn to recognize these traits.

Yet, even in recognizing them it doesn't mean we're comfortable with them. Most people would like to remove them. They'd like to remove them because they have no consistent access. So in looking at somebody's design, we're entering into a very, very special place, a place in which the true self can really be revealed.

When I first began doing rave charts in 1987, because of the fact that the design was red, and the personality was black, and that I was coloring in all those charts, there was no computer program, it was necessary to color in the design first. And it was always a revelation, a deep insight in many ways into the nature of somebody's unconscious as a whole.

Here is a classic example. We have somebody that at the person-

> In looking at this illustration of the design, one can see that every single one of those gates represents a mystery and a surprise in this being's life.

ality level has strong definition and is a split definition generator. And yet when we're looking at the design, you can see that in the context of the design, this being is a reflector. In other words, there is no strongly fixed aspect at the unconscious level. This is a highly vulnerable unconscious in that sense. And here is somebody that without a fixed nature in the unconscious doesn't have a stable consistent way in which their unconscious is going to operate.

The other thing is the mystery of what juxtaposition is, the mystery of the human binary, of the DNA and RNA, the mystery of the whole always being greater than the sum of its parts. When

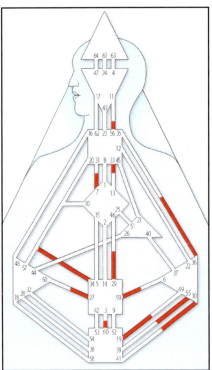

we take the generator split definition personality and we combine it with the reflector design, we end up with an emotional split definition generator and the emotional system suddenly emerges out of nowhere. The unconscious, there is no consistent emotional nature. There is none in the personality. Yet when they come together as a being, when the totality is there, suddenly that is something that emerges. And it's so important to see that there is a great divide between the personality and the design, the divide of identification.

Here looking at this illustration of the design, one can

see that every single one of those gates represents a mystery and a surprise in this being's life. And every single one of those gates represents an aspect that they themselves do not understand why they are there and why they function the way they do. As I said, they learn to recognize them as traits, but that does not mean that they understand them. It does not mean they are secure in them. It does not mean any of that.

Let's take some examples. We have in the throat center, the 56th gate, and the 56th gate is a gate of stimulation. It is a gate that is there to be able to tell stories, to stimulate other people, to get them interested in ideas. It is a gate that is deeply, deeply seduced by ideas in the world. It is also a voice gate; that is, it is a throat gate, it has a voice. And all the 11 gates in the throat have a voice. It's so important for somebody to know their own unique voice.

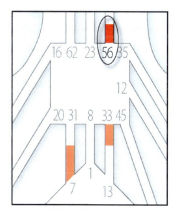

But think about what that's like. Here's somebody that has two unconscious gates in the throat. They don't know where those voices come from. Not only do they not know where they come from, but remember they don't identify with them. So here's the 56th gate, and what the 56th gate says is, "I believe or I don't believe." Now, imagine what that's like, you're the passenger, you're the personality, you're not identified with that voice. You're in a conversation with somebody, they say that there's blue cheese on Mars, and there's this unconscious 56 going yes, I believe that's so. Now, I'm being a bit absurd; nonetheless, please understand the dilemma that comes out of that. The dilemma that comes out of that is "do I really believe that? Why did I say that I believe that if I don't know if I believe that?"

It's like the 33rd gate in the throat. Here is the gate of the Retreat, one of the four gates of aloneness. And so here is a theme

19

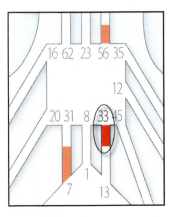

of aloneness in tune with the personality second line hermit and its need for withdrawal. Yet, the 33rd gate is a gate of remembrance. It has the capacity to be able to remember things. Think about what it's like for this person in relationship to their memory system. Their personality doesn't identify with having a good memory because the personality isn't strong in memory gates. But here's a very powerful memory gate, and all of a sudden things pop out of nowhere.

This person doesn't respond to somebody's inquiry by saying, yes I've got a strong memory, and I've got good access to it, or whatever. What they end up saying is, "I don't know, sometimes I remember things, sometimes I don't." The fact of the matter is that that process is always going on. But it's always going on at the unconscious level so they do not have this capacity to recognize it, nor recognize why they suddenly remember something, or why they recognize it and remember it and speak of it in the way in which they do. All of that's operating from the unconscious. All of that is coming out of the tunnel.

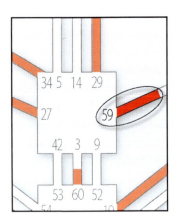

One of the major themes in an individual reading for this being would be, of course, the focus on the nature of the emotional system and how the emotional system operates. Here is somebody because of the unconscious 59th gate, that does not have clear access to or identification with their sexual strategies. The 59th gate, the gate of Dispersion, this is the gate of our roles in terms of the way in which our genetics operate, the way in which we are going to try to find the right being to bond with in order to be able to reproduce, in order to be able to be successful in intimacy.

This strategy, the genetic strategy of the 59th gate is all operating at the unconscious level. So this is not somebody who understands why they approach intimacy in the way they do. As a matter of fact, given their personality is a second line, and that it's resistance in that sense, it would like to be like to be left alone, it is a line of shyness. It is about withdrawal despite the fact that it's open to the call. That not being able to identify with one's sexual strategy one is more comfortable, in that sense, being left alone.

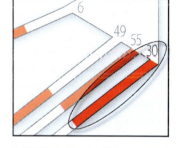

If we look at the solar plexus system itself, we will see that there are three unconscious gates that are activated there. And two of those gates, the 30th gate which is very important because it's part of the profile, it's the design Earth in the fourth line, that fourth line of burnout, of fatigue. That 30th gate is the gate of feelings; it's the gate of desire, the Clinging Fire. And at the other side in the solar plexus system, in the same stream, this stream of feelings that begins in the 41st gate, goes to the 30, and then from the 36 to the 35 in the throat. This is the human experiential way. And the human experiential way and its potential is in the 36th gate, the Darkening of the Light, the potential for crisis.

But this is also about desire. This stream is highly sexual. And the 30th gate and the 36th gate are deeply sexual in the way in which they need

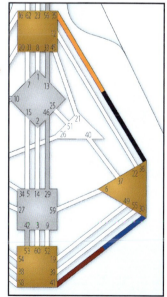

Stream of Feelings

to fulfill themselves through experience in this life. Yet, desire drives them. And that there is a need within them in order for them to have progress to be able to satisfy those desires and be able to move on to new desire experiences.

Think about what it's like for this being. That whole desire system is operating in the emotional wave. There are moments that this being wakes up and they are deeply desirous. Deeply desirous of a being, of a thing, of an experience, whatever the case may be. And yet, that desire is not something they can identify with. They don't know why they're being pulled that way. They don't know why they're being pulled to that person, or that experience, or that thing.

Not only that. They don't know why they're either up or down in relationship to it. The hope end of the wave in the motor system, that hope, when it's attached to desire, can be something that's enormously exciting. You know, the possibilities, the potential, the deep, deep expectation that comes with this stream.

> The 30th gate mystically is the gate of the fates.

Yet, if you're not identified with the process, that kind of force within you can feel very uncomfortable. The 30th gate mystically is the gate of the fates. It's the place in which we learn to recognize that despite everything, there are forces greater than us. And that in the end, any expectation we have, no matter how well-prepared we are, no matter how gifted, no matter how fortunate, that there are always forces that can get in the way.

Here is somebody that has to experience the pressure of desire on them, and a pressure of desire that's always operating. After all, in the integrated design, the whole emotional system is pulsating through and connected to the sacral, it is part of their process. And in not living out their nature, part of their deep frustration—"Why do I want that? Why don't I want this? Why am I being pulled in that direction?"

Also, in the solar plexus system is the 55th gate. The 55th gate is important in the sense that it resonates deeply with the strong individuality of the personality, strong individual definition that is there within this design. And the 55th gate is a melancholic gate. It is the most melancholic in the sense that it is the moodiest of gates.

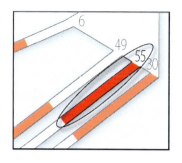

But in the mundane context of analysis, the 55th gate ultimately is about indecisiveness. Not being able to make a decision about love, about friendship, about career, about this, about that. And not being able to make that decision simply because the knowing isn't there in the moment.

Individuals can only act when they know. This is the nature of the pulse; this is the nature of the mutative process. And that that knowing can wait awhile before it happens. But it places these beings under a lot of pressure from other people that have an expectation that they should know in that moment what to decide and what not to decide.

> Individuals can only act when they know.

In the context of this being's complete design, this can be a gate that can be a great advantage to explore. In other words, it resonates to the emotional authority that this being has to trust in its life. Without trusting its emotional authority, without being indecisive, without being able to wait for the wave, and wait for time, only then are they going to be clear.

And at the moment that they understand that what is their basic indecisiveness is not something to overcome. Remember, they don't identify with the 55. It's not something for them to overcome at all. It's something for them to embrace. To recognize that clarity has a moment for them, clarity will come. But they have to wait for that clarity to be there. Only then can they act

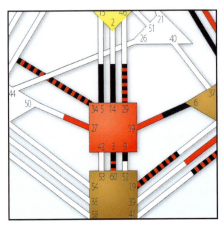

and can they act in a healthy manner.

Look at the configuration in the sacral center. It's very interesting. Whenever we have activation from the design side and the personality side in the same gate, obviously we have the ability within the personality to be able to identify with the general process of that gate. In other words, it's less of a mystery, obviously and less of a surprise so there are nuances that exist when there are these conjunctions between design and personality in a gate.

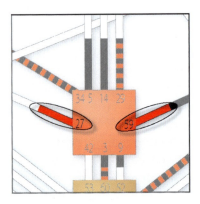

But in the context in which we're examining somebody's design, very important to see that the real surprises and the real mystery, that is entirely in the 59 and the 27 in this design. In other words, those are going to be the aspects of the way in which the sacral center operates that are going to be the most confusing to this person. And of course, because of the situation in which the 59 is opening up to such a powerful emotional system and all the activation that is there, and that the 27 is opening up to an extremely vulnerable and entirely open, no activation at all, splenic center, those two gates are going to distort, in a sense, the clarity that could come through the personality in understanding the way in which their sacral operates. In other words, there are real aspects of mystery for them in the way in which these two gates are going to function.

Both the 27th gate and the 59th gate are role gates. That is, they're gates of our genetic strategies, the genetic strategy for bonding that's there in the 59th gate, and the genetic strategy for caring that's in the 27th gate. It's part of the defense circuit; it's part of a basic tribal nature in us. And here we're dealing with somebody who is fundamentally individual, and so they don't necessarily resonate to the tribal aspects in their design.

The 59th gate and the 27th gate are both second lines. In other words, they both represent themes that are in resonance with the personality. That is, the personality profile has the second line theme of the hermit, the second line theme of shyness. Now, think about the difficulty for this human being. We've already seen that at the emotional level, at the unconscious level in the tunnel there is all this desire, all this need for sexual fulfillment, all this need for sexual experience. And through the 55th gate, difficulty in that sense, fickleness about who to love and who not to love and so forth and so on.

Yet, think about the 59.2, this is fundamentally a line of shyness. Here is somebody who is both driven by desire and not allowed, in that sense, by their strategy to go after anything, to be restrained in that. And to hold all of that desire, frustrating, hold all of that desire within them. One of the things to recognize about genetic role gates is that they're strategies, and they're strategies that are there in order to attract exactly the right bonding partner. In the case of the 59.2, the 59.2 is there to be reached by boldness, that their barriers are broken down. And that only when the barrier can be broken down are they going to be open and accessible. It does not mean that they are a shy person. It means it's the way in which they operate genetically; they need to bring powerful forces to them that will call them. Again, a theme in this being's life.

You see, the overall person, the whole design, if they're following and honoring their type, and they're following the nature of their strategy, if they're honoring their authority, here is an emotional generator. To have the right connection with somebody begins with being called. Not simply that it's a matter of their profile, and that costume that fits them best, but they are a generator, and it is natural for them to respond, to respond to the call. And yet, they have to see that being an emotional being means that responding to the call means that one has to be necessarily reticent in the moment.

The spontaneous is fundamentally dangerous for this being. And that in waiting out the wave and making it difficult for the other to get access to them, playing hard to get, that they're fulfilling their genetic strategy, that it will bring to them the right person into their life, the right contact into their life. Again, in design we have what's called genetic continuity. There is a fundamental relationship between all aspects together in the whole. It is correct for this being to wait. It is correct for them to wait to respond to be called. It is correct for them to be indecisive. It's right for them so that they can go through their emotional process, so they can be clear.

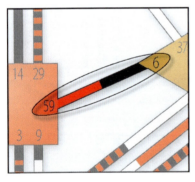

And the 59/6 as a channel, this channel of intimacy, does not display powerful emotion regularly. It's not as volatile emotionally as the other streams. It's more subtle in the way in which it deals with its emotional energy, and here is somebody without access. So it's so important to see that no matter how far you can go in describing the mechanical variance that is there in the design, in the end the only healing for the being is if they experiment with following their strategy. Following a strategy, all the pieces fit together because all the pieces fit together in the continuity.

It's the same with the 27th gate, the gate of Nourishment, the gate of caring. It's so important for this being because it's one of the two ways in which they can activate their splenic system. Here is somebody with a totally open spleen who can be without, literately without fear, fearless to the point of foolishness. And here is somebody because the spleen is open, in which fear is a theme in their life. And most of all, the fear of taking responsibility.

For this being, to care about somebody else, to care for something else, to take responsibility for something else, this can only come through being called. It is the only way. It's a second line theme. And it goes back to the fundamental nature of what this person is. They are here to be called. And they're here to wait out the wave.

Roads and tunnels, that's the starting place. When we come together and we bring the design and the personality and integrate it into the body graph, we do get the being. It's important to see that in understanding the way in which these two aspects operate, the way in which we identify it through the personality and are surprised through the design, that you have to be careful in recognizing that this is an insight into infrastructure. But it's only when you see the whole together that you really get the sense of the being. And you also get to understand how the genetic continuity is working, whether it's on the unconscious side or the personality side.

In the end, what we're dealing with is a split definition emotional generator. And whether or not this being has access or not, the theme for them in their process is going to remain the same, they have to be able to honor the mechanics of their design. And in honoring the mechanics of their design, then they're going to be able to see that it's possible for them to eliminate resistance, and that is the secret of this knowledge.

We have difficulty within ourselves because of our discomfort with the unconscious. Human Design provides that opportunity to be able to see mechanically the way in which the so-called unconscious is operating in us. But it still doesn't mean that gives us access. It doesn't. It means that intellectually that it can be stimulating for us, that it can give us an insight that can resonate within us, and ultimately impact on the nature of our overall awareness.

But the adult human being is deeply conditioned. And it's not enough to simply understand a process. And it's not enough to be able to agree intellectually in understanding a process. In the end, everything having to do with Human Design is the practice of a human design. And that practice of Human Design is truly simple. It is about honoring your type and its strategy.

The following of that strategy for this being integrates all of that in a healthy way in their life. By learning to respond, by learning to be able to say "wait, I need to be clear," by giving themselves a chance, timewise to go through a process in which they move emotionally and they get to see the clarity at the end, all of that is going to bring them the reward of being able to make healthy decisions.

Nothing is more important in this life. This is somebody that can be very confused at the mental level, it's all open. They can be very fearful and very uncomfortable. They have an undefined ego; they can undervalue themselves and not see that they are worthy. For this kind of a human being, driven by that kind of openness and conditioning dealing with a split definition in which their throat is often saying things they cannot do and will not do, that it's only the simple mechanic that's going to liberate them and it's going to take time.

It is so important to understand that about the adult and deconditioning. It takes time. It takes seven biological years. It takes that amount of time to change the vast majority of the cells in the body; it's a deconditioning process to move from the not-self to the true self. The knowledge and insight that we have is beautiful, it's wonderful. However, it's the practice that brings awakening. Each day that one practices and works at one's strategy, doesn't mean you have to be perfect, and it doesn't mean you always have to do it. As a matter of fact, the greatest learning process with this knowledge is ignoring your strategy to see the price that you pay; in order for your awareness to grasp deeply, that following your strategy will be a boon. And it is.

So in looking at these various aspects, they're so important for you in understanding what's there in a being, what's there in yourself. But recognize that once you have that knowledge and you have that information, it's through the practice of Human Design that you get the affirmation, that you see that what is intellectually described, functions in fact. And then you can take the burden off your shoulders and see that it is not necessary for you be concerned about where this is coming from and where that is coming from. All of it in the end is you, and you operating correctly in the wave.

> It's the practice that brings awakening.

Printed by Amazon Italia Logistica S.r.l.
Torrazza Piemonte (TO), Italy